Index

1. What is Diesel Mechanics..3
2. Diesel Engine...3
3. Main Components..5
4. Working Principal of a Diesel Engine.............................9
5. Advantages of a Diesel Engine..10
6. Commonly Repaired Problems on Diesel Engines.......11
7. Diesel Engine Troubleshooting.......................................14
8. Basic Diesel Engine Troubleshooting Chart..................14
9. Precautions for maintenance a Diesel Engine..............18
10. Tips for maintain a Diesel Engine..................................19
11. Detailed Summary..21

Start of Book

What is Diesel Mechanics:

Diesel Mechanics. Diesel mechanics inspect and repair vehicles with diesel engines, including buses, trucks and construction machines. To repair these vehicles and diagnose malfunctions, they inspect parts and systems, check batteries and adjust wheels. Being a diesel mechanic can be physically demanding.

Diesel Engine:

A type of internal combustion engine which ignites fuel by injecting it into hot, high-pressure air in a combustion chamber. It has neither carburettor nor ignition system. The fuel is injected in the form of a very fine spray, by means of a nozzle, into the combustion chamber. There it is ignited by the heat of compressed air which the chamber has been charged with. The diesel engine operates within a fixed sequence of events, which may be achieved either in four or two strokes. The low-speed (i.e. 70 to 120 rev/min) two-stroke diesel is used for main propulsion units, since it can be directly coupled to the propeller and shafting. The medium speed four-stroke engine (250 – 1200 rev/min) is used for the auxiliaries such as alternators and also for main propulsion with a gearbox.

A four-stroke diesel engine resembles a gasoline engine as it works on the four-stroke cycle, that is: admission, compression, power and exhaust. When the piston gets down on the air admission

stroke, the lower pressure in the cylinder allows a charge of air into the cylinder through the inlet valve which opens just before top dead centre.

Once the piston has passed the bottom dead centre and is beginning to ascend, the inlet valve closes and the upward movement of the piston compresses the air charge in the cylinder causing a quick rise of temperature. Before the second stroke is over, the charge of fuel oil is gradually injected into the cylinder by an injector.

The burning of the air-fuel charge makes the gases expand. They push the piston downwards and create the power stroke. Before the piston reached the bottom dead centre, the exhaust valve opens and, as the piston goes up again, the burnt gases are forced out through the exhaust valve. Just before top dead centre the inlet valve opens and the cycle begins again.

- **High-speed diesel engine** – Trunk piston type engine having a rated speed of 1400 rpm or above.

- **Medium-speed diesel engine** – Trunk piston type engine with speed range from 400 rpm to 1200 rpm.

- **Low-speed diesel engine** – Crosshead type engine with rated speed of less than

400 rpm.

Main components:

1. Engine block

The engine block is made of nodular cast iron in one piece for all cylinder numbers. The main bearing caps are fixed from below by two hydraulically tensioned screws. They are guided sideways by the engine block at the top as well as at the bottom. Hydraulically tensioned horizontal side screws support the main bearing caps.

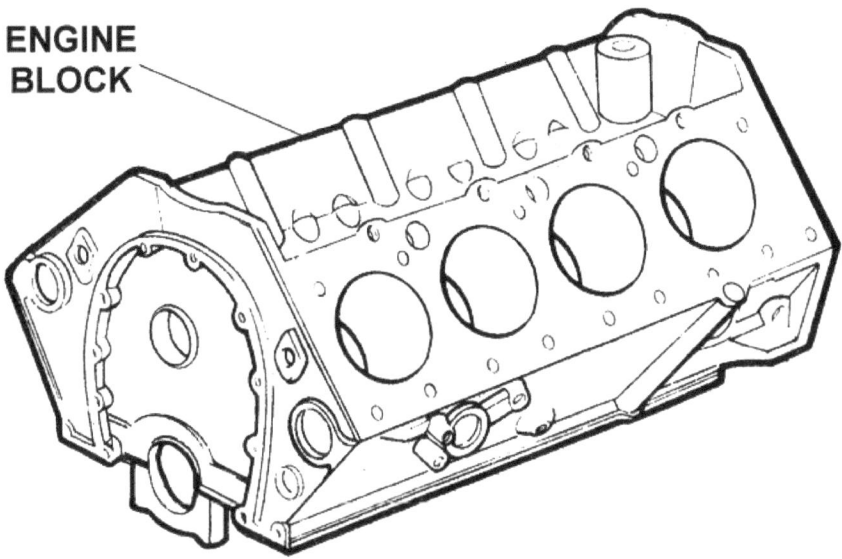

2: Crankshaft

The crankshaft is forged in one piece. Counterweights are fitted on every web. High degree of balancing results in an even and thick oil film for all bearings.

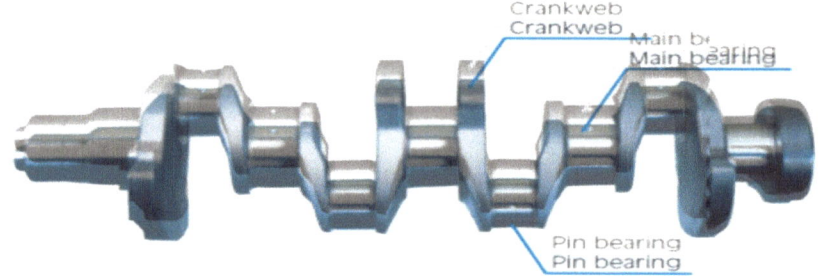

3: Connecting rod

The connecting rod of alloy steel is forged and machined with round sections. The lower end is split horizontally to allow removal of piston and connecting rod through the cylinder liner. All connecting rod bolts are hydraulically tightened. The gudgeoning bearing is of tri-metal type. Oil is led to the gudgeon pin bearing and to the piston through a bore in the connecting rod.

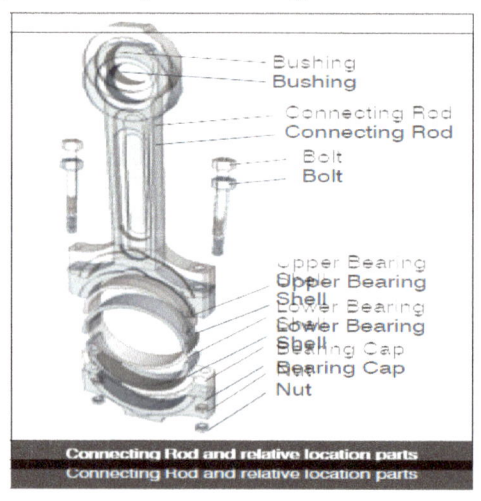

4: Main bearings and big end bearings

The big end bearings are of tri-metal type with steel back, lead bronze lining and a soft and thick running layer. Both tri-metal and bi-metal bearings are used as main bearings.

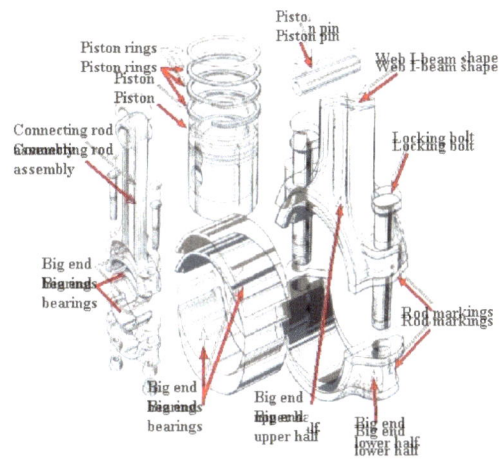

5: Cylinder liner

The centrifugally cast cylinder liner has a high and rigid collar to minimise deformations. The liner material is a special grey cast iron alloy developed for excellent wear resistance and high strength. Accurate temperature control is achieved with precisely positioned longitudinal cooling water bores. To eliminate the risk of bore polishing, the liner is equipped with an anti-polishing ring. The cooling water space between block and liner is sealed off by double O-rings. In the upper end the liner is equipped with an anti-polishing ring to eliminate bore polishing and reduce lube oil consumption.

6. Piston and piston rings

The piston is of composite design with nodular cast iron skirt and steel crown. The piston skirt is pressure lubricated, which ensures a controlled oil distribution to the cylinder liner under all operating conditions. Oil is fed to cooling gallery in the piston top through the connecting rod. The piston ring grooves are hardened for good wear resistance. The piston ring set consists of two directional compression rings and one spring-loaded conformable oil scraper ring. All piston rings have a wear resistant chromium plating.

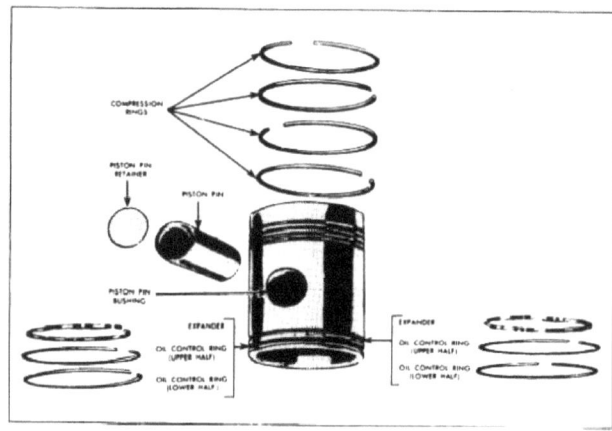

7. Cylinder head

The cylinder head is designed for easy maintenance with only four hydraulically tightened studs. No valve cages are used, which results in very good flow dynamics in the exhaust gas channel. The exhaust valve seats are water cooled and all valves are equipped with valve rotators. The seat faces of the inlet valves are Stellite-plated. In case the engine is specified for MDF operation only, also the exhaust valves are Stellite-plated. Engines that are intended for operation on HFO have Mnemonic exhaust valves

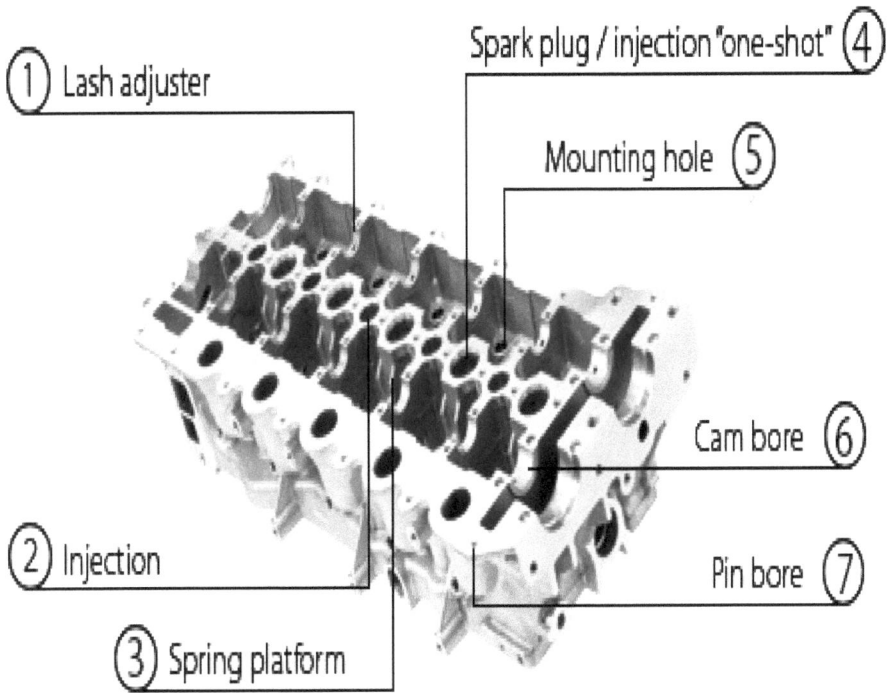

Working Principal of a Diesel Engine:

Like a gasoline engine, a diesel engine is a type of internal combustion engine. Combustion is another word for burning, and internal means inside, so an internal combustion engine is simply one where the fuel is burned inside the main part of the engine (the cylinders) where power is produced. That's very different from an *external* combustion engine such as those used by old-fashioned steam locomotives. In a steam engine, there's a big fire at one end of a boiler that heats water to make steam. The steam flows down long tubes to a cylinder at the opposite end of the boiler where it pushes a piston back and forth to move the wheels. This is external combustion because the fire is *outside* the cylinder (indeed, typically 6-7 meters or 20-30ft away). In a gasoline or diesel engine, the fuel burns inside the cylinders themselves. Internal combustion wastes much less energy because the heat doesn't have to flow from where it's produced into the cylinder: everything happens in the same place. That's why internal combustion engines are more efficient than external combustion engines (they produce more energy from the same volume of fuel):

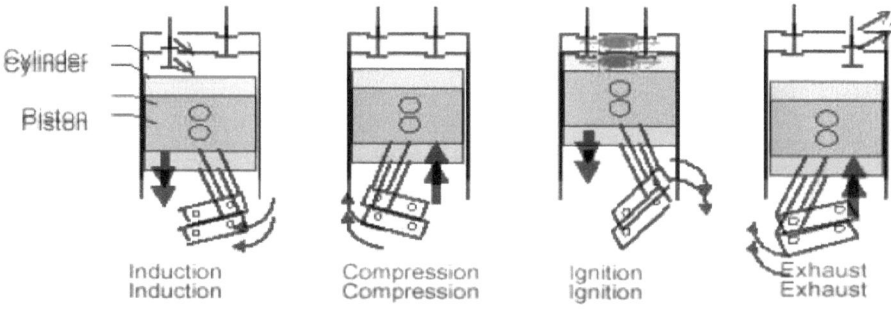

Advantages of a Diesel Engine:

The diesel engine is much more efficient and preferable as compared with gasoline engine due to the following reasons:

- Modern diesel engines have overcome disadvantages of earlier models of higher noise and maintenance costs. They are now quiet and require less maintenance as compared with gas engines of similar size.
- They are more rugged and reliable.
- There is no sparking as the fuel auto-ignites. The absence of spark plugs or spark wires lowers maintenance costs.
- Fuel cost per Kilowatt produced is thirty to fifty percent lower than that of gas engines.
- An 1800 rpm water cooled diesel unit operates for 12,000 to 30,000 hours before any major maintenance is necessary. An 1800 rpm water cooled gas unit usually operates for 6000-10,000 hours before it needs servicing.
- Gas units burn hotter than diesel units, and hence they have a significantly shorter life compared with diesel units.

Applications & Uses for Diesel Engines

Diesel engines are commonly used as mechanical engines, power generators and in mobile drives. They find wide spread use in locomotives, construction equipment, automobiles, and countless industrial applications. Their realm extends to almost all industries and can be observed on a daily basis if you were to look under the hood of everything you pass by. Industrial diesel engines and diesel powered generators have construction, marine, mining, hospital, forestry, telecommunications, underground, and agricultural applications, just to name a few. Power generation for prime or standby backup power is the major application of today's diesel generators. Check out our article on the various types of engines and generators and their common applications for more examples.

Power Generators:

Diesel powered generators, or electrical generator sets, are used in countless industrial and commercial establishments. The generators can be used for small loads, such as in homes, as well as for larger loads like industrial plants, hospitals, and commercial buildings. They can either be prime power sources or standby/back-up power sources. They are available in various specifications and sizes. Diesel generator sets rating 5-30KW are typically used in simple home and personal applications like recreational vehicles. Industrial applications cover a wider spectrum of power ratings (from 30 kW to 6 Megawatts) and are used in numerous industries throughout the globe. For home use, single-phase power generators are sufficient. Three-phase power generators are primarily used for industrial purposes.

Commonly Repaired Problems on Diesel Engines:

Right now is a great time to own a truck. With diesel prices down across the board, it makes it a more economical choice, and the extra power isn't a bad thing either. It's important to note, however, that diesel trucks aren't invincible. Like every vehicle, they come with their share of problems. If a defective engine isn't properly serviced, it can cause internal combustion and potentially destroy your entire machine. Here are some of the most common diesel truck repairs to watch for.

1. **Oxidized Oil**

Diesel trucks that sit in one place too long, operate infrequently, or remain in storage between seasons often have problems with the oil oxidizing. In other words, air gets into the oil, and creates bubbles that can interfere with proper lubrication, which can result in a faltering or even damaged engine. Even though the oil isn't technically dirty, it needs to be changed as soon as possible after this idle period.

2. Humidity Reactions

Water is another element that can contaminate the lubricant in the engine and cause adverse reactions. If a truck sits for too long or idles for an extended period of time in a humid or precipitous area, the hydration can cause the engine to knock. Water attacks additives and increases oxidation. It can also interfere with the lubrication process, which can lead to a severely damaged machine.

3. Black Exhaust

If you've driven behind a diesel truck, you know that they generally exhibit more smoke than traditional vehicles. They can also release a highly unpleasant odour that can stink up the cab and make it difficult to breathe. Not to mention, you'll likely be slapped with a hefty fine for ignoring the clean air ordinance in your state.
The exhaust is generally a result of an imbalanced air to fuel ratio, leaning on the side of too much fuel and not enough air. A faulty injector, injector pump, air filter, EGR valve, or even turbocharger could be the root of the problem.

4. Hard Starting

Some diesel engines have a difficult time starting or they experience a delayed start. This is usually a sign of low compression or a fuel delivery issue. Some diesel engines simply crank a little when you start them, which is perfectly normal. But if it has an extremely difficult time starting, cranks more than normal, or won't start at all, it's important to get things checked out as soon as possible.

5. Lack of Power

Another fuel related problem manifests itself within a lack of power. You'll notice this when it has issues starting or accelerating. Dirty fuel filters, loose throttle linkage, excessive lubrication, and issues with the fuel injectors can lead to this issue.

6: Failing Lead-Acid Storage Batteries

There's often a heavy load on the lead-acid storage battery, which is a useful component in the engine's starter system. If the storage battery malfunctions or doesn't work properly, it can cause an imbalanced compression ratio, which can influence the starter system negatively.

7: Defective Glow Plug

Diesel engines don't have spark plugs to ignite the fuel-air mixture in their cylinders like gasoline-powered cars. They rely on glow plugs to ignite the mixtures through a high-resistance heating element, similar to the elements in a stove coil or toaster. When the glow plug goes bad, it makes it nearly impossible for the engine to start, particularly in cold weather.

8: Contaminated Fuel

Because diesel is much more viscous than gasoline, it can become more easily contaminated. The four most common, and equally dangerous, fuel contaminants include glycol, dilution, soot, and water. If any of these contaminants penetrate the fuel system, it can lead to major engine disruption.

9: Higher Compression Ratio

The average diesel engine has a compression ratio of 20:1, while the average gas powered engine has an average ration of 8:1. This high compression ratio makes the engine more powerful, smooth, and sometimes more efficient, but it can also lead to issues. For example, it can cause the engine to knock more often as a result of an undesirable burn pattern, and can also contribute significantly to fuel injection problems.

10. Noise

Nose from a diesel engine can be significant, and can also be a sign of something wrong. Diesel engines are naturally louder than other vehicles, but if you notice inconsistent noise or distinct knocking in the engine, it could be a sign of a problem with the fuel injectors, which can affect the compression balance and reduce performance.

11. Wrong Weight Viscosity

Hard starting is often the result of the wrong weight viscosity of engine lubrication. Viscosity in diesel lubricants is much higher than in gasoline, and many people often get the viscosity weight wrong when replacing the oil. Other times, they'll simply use a single-weight engine oil during hot weather and then forget to switch back to multi-viscosity oil when the weather turns cold. It's best to simply use a multi-viscosity oil year-round to avoid forgetting to switch and risking damaging the engine.

When diagnosing the problem with your diesel truck, it's best to look at the basics first in order to save time and energy. Operators should take good care of their trucks, perform routine maintenance to avoid the worst issues, and always report engine troubles to a trusted mechanic who specializes in the care of diesel engines.

If you're looking for a trusted mechanic, look no further than RC Auto Specialists in Tulsa. We're your one stop shop for anything diesel vehicle repair related. With 80 combined years of experience, our experts use the latest tools and knowledge available to diagnose the problem and get your diesel truck back on the road where it belongs. To schedule an appointment, contact us today!

DIESEL ENGINE TROUBLESHOOTING:

NOTE: This is GENERAL information. This article is not intended to be specific to any unique situation or individual vehicle configuration. The purpose of this Troubleshooting information is to provide a list of common causes to problem symptoms. For model-specific Troubleshooting, refer to SUBJECT, DIAGNOSTIC, or TESTING articles available in the section(s) you are accessing.

NOTE: Diesel engines mechanical diagnosis is the same as gasoline engines for items such as noisy valves, bearings, pistons, etc. The following trouble shooting covers only items pertaining to diesel engines.

BASIC DIESEL ENGINE TROUBLESHOOTING CHART:

CONDITION & POSSIBLE CAUSE	CORRECTION
Engine Won't Crank	
Bad battery connections or dead batteries	Check connections and/or replace batteries
Bad starter connections or bad starter	Check connections and/or replace starter
Engine Cranks Slowly, Won't Start	
Bad battery connections or dead batteries	Check connections and/or replace batteries
Engine oil too heavy	Replace engine oil
Engine Cranks Normally, But Will Not Start	
Glow plugs not functioning	Check glow plug system, see FUEL SYSTEMS
Glow plug control not functioning	Check controller, see FUEL SYSTEMS

CONDITION & POSSIBLE CAUSE	CORRECTION
Fuel not injected into cylinders	Check fuel injectors, see FUEL SYSTEMS
No fuel to injection pump	Check fuel delivery system
Fuel filter blocked	Replace fuel filter
Fuel tank filter blocked	Replace fuel tank filter
Fuel pump not operating	Check pump operation and/or replace pump
Fuel return system blocked	Inspect system and remove restriction
No voltage to fuel solenoid	Check solenoid and connections
Incorrect or contaminated fuel	Replace fuel
Incorrect injection pump timing	Re-adjust pump timing, see FUEL SYSTEMS
Low compression	Check valves, pistons, rings, see ENGINES
Injection pump malfunction	Inspect and/or replace injection pump
Engine Starts, Won't Idle	
Incorrect slow idle adjustment	Reset idle adjustment, see TUNE-UP
Fast idle solenoid malfunctioning	Check solenoid and connections
Fuel return system blocked	Check system and remove restrictions
Glow plugs go off too soon	See glow plug diagnosis in FUEL SYSTEMS
Injection pump timing incorrect	Reset pump timing, see FUEL SYSTEMS
No fuel to injection pump	Check fuel delivery system
Incorrect or contaminated fuel	Replace fuel
Low compression	Check valves, piston, rings, see ENGINES

CONDITION & POSSIBLE CAUSE	CORRECTION
Injection pump malfunction	Replace injection pump, see FUEL SYSTEMS
Fuel solenoid closes in RUN position	Check solenoid and connections
Engine Starts/Idles Rough W/out Smoke or Noise	
Incorrect slow idle adjustment	Reset slow idle, see TUNE-UP
Injection line fuel leaks	Check lines and connections
Fuel return system blocked	Check lines and connections
Air in fuel system	Bleed air from system
Incorrect or contaminated fuel	Replace fuel
Injector nozzle malfunction	Check nozzles, see FUEL SYSTEMS
Engine Starts and Idles Rough W/out Smoke or Noise, But Clears After Warm-Up	
Injection pump timing incorrect	Reset pump timing, see FUEL SYSTEMS
Engine not fully broken in	Put more miles on engine
Air in system	Bleed air from system
Injector nozzle malfunction	Check nozzles, see FUEL SYSTEMS
Engine Idles Correctly, Misfires Above Idle	
Blocked fuel filter	Replace fuel filter
Injection pump timing incorrect	Reset pump timing, see FUEL SYSTEMS
Incorrect or contaminated fuel	Replace fuel
Engine Won't Return To Idle	
Fast idle adjustment incorrect	Reset fast idle, see TUNE-UP
Internal injection pump malfunction	Replace injection pump, see FUEL SYSTEMS
External linkage binding	Check linkage and remove binding
Fuel Leaks On Ground	

CONDITION & POSSIBLE CAUSE	CORRECTION
Loose or broken fuel line	Check lines and connections
Internal injection pump seal leak	Replace injection pump, see FUEL SYSTEMS
Cylinder Knocking Noise	
Injector nozzles sticking open	Test injectors, see FUEL SYSTEMS
Very low nozzle opening pressure	Test injectors and/or replace
Loss of Engine Power	
Restricted air intake	Remove restriction
EGR valve malfunction	Replace EGR valve
Blocked or damaged exhaust system	Remove restriction and/or replace components
Blocked fuel tank filter	Replace filter
Restricted fuel filter	Remove restriction and/or replace filter
Block vent in gas cap	Remove restriction and/or replace cap
Tank-to-injection pump fuel supply blocked	Check fuel lines and connections
Blocked fuel return system	Remove restriction
Incorrect or contaminated fuel	Replace fuel
Blocked injector nozzles	Check nozzle for blockage, see FUEL SYSTEMS
Low compression	Check valves, rings, pistons, see ENGINES
Loud Engine Noise With Black Smoke	
Basic timing incorrect	Reset timing, see FUEL SYSTEMS
EGR valve malfunction	Replace EGR valve
Internal injection pump malfunction	Replace injection pump, see FUEL SYSTEMS

CONDITION & POSSIBLE CAUSE	CORRECTION
Incorrect injector pump housing pressure	Check pressure, see FUEL SYSTEMS
Engine Overheating	
Cooling system leaks	Check cooling system and repair leaks
Belt slipping or damaged	Check tension and/or replace belt
Thermostat stuck closed	Remove and replace thermostat, see ENGINE COOLING
Head gasket leaking	Replace head gasket
Oil Light on at Idle	
Low oil pump pressure	Check oil pump operation, see ENGINES
Oil cooler or line restricted	Remove restriction and/or replace cooler
Engine Won't Shut Off	
Injector pump fuel solenoid does not return fuel valve to OFF position	Remove and check solenoid and replace if needed

VACUUM PUMP DIAGNOSIS

CONDITION & POSSIBLE CAUSE	CORRECTION
Excessive Noise	
Loose pump-to-drive assembly screws	Tighten screws
Loose tube on pump assembly	Tighten tube
Valves not functioning properly	Replace valves
Oil Leakage	
Loose end plug	Tighten end plug
Bad seal crimp	Remove and re-crimp seal

Precautions for maintenance a Diesel Engine:

You must obtain the specific safety pre-cautions for a given engine from the appropriate NAVSEA technical manual. In addition to the guidelines in the NAVSEA technical manual, you should observe the following precautions when operating or maintaining a diesel engine.

Relief Valves

If a relief valve on an engine cylinder lifts (pops) several times, stop the engine immediately. Determine the cause of the trouble and decide upon the correct solution. Except in an emergency, NEVER lock a relief valve in the closed position. Pressure-relief mechanisms are fitted on enclosures in which excessive pressures may develop.

Fuel

When fuel reaches the injection system, it should be absolutely free of water and foreign matter. You must thoroughly centrifuge the fuel before using it, and you must keep the filters clean and intact. Remember, fuel leakage into the lubricating oil system will cause dilution of the lubricating oil with a consequent reduction in viscosity and lubricating properties.

Cooling Water

Do NOT allow a large amount of cold water, under any circumstances, to enter a hot engine suddenly. Rapid cooling may crack a cylinder liner and head or may cause a piston to seize within a cylinder. Reduce the load or, when ordered to do so, stop the engine when the volume of circulating water cannot be increased and the temperatures are too high. In freezing weather, you must carefully drain all passages and pockets in the engine that contain fresh water and that are subject to freezing, unless an antifreeze solution has been added to the water.

Starting Air

When engines are stopped, you must vent all starting-air lines. Serious accidents may result if pressure is left on. Intake air must be kept as clean as possible. Accordingly, you must keep all air ducts and passages clean.

Cleanliness

Cleanliness is essential to efficient operation and maintenance of diesel engines. You must maintain clean fuel, clean coolants, clean lubricants, and a clear exhaust. You must also keep the engines clean at all times, and take steps to prevent oil or fuel from accumulating in the bilges or in other areas to prevent fire hazards.

Tips for maintain a Diesel Engine:

While most veteran professional mechanics are trained in safe working practices, service technicians new to the equipment rental field must often learn by doing. Therefore, mechanics new to the industry should be well-versed in the following basic safety precautions:

1.) Always ensure that the engine is properly supported and in safe condition before you attempt to use force to loosen any nuts, bolts or plugs. Wherever possible, initially slacken tight fastenings before raising the vehicle off the ground or removing the engine from its mountings.

2.) A vehicle should always have its gearbox in neutral; never start the engine unless the load has been removed and the hand brake applied.

3.) Never run catalytic converter-equipped engines without the exhaust system heat shields in place.

4.) Before starting work, allow oil, coolant or other fluids to cool to avoid scalding.

5.) Never siphon fuel, coolant, cooling agents, solvents or liquids by mouth or allow prolonged contact with skin.

6.) Take off watches, rings or other jewelry. Keep long hair and loose clothing well out of the way of any moving parts.

7.) Keep tools away from the top of the battery. They could cause a short circuit and possible explosion.

8.) Never lean over to work on a running engine.

9.) Never take risky shortcuts or rush to finish a job.

10.) Don't pour drained engine fluids down the drain. Don't mix used oil with other materials, such as paints and solvents. Take used oil to an oil-recycling bank.

Detailed Summary:

Diesel engine, any internal-combustion engine in which air is compressed to a sufficiently high temperature to ignite diesel fuel injected into the cylinder, where combustion and expansion actuate a piston. It converts the chemical energy stored in the fuel into mechanical energy, which can be used to power freight trucks, large tractors, locomotives, and marine vessels. A limited number of automobiles also are diesel-powered, as are some electric-power generator sets.

Diesel Consumption:

The diesel engine is an intermittent-combustion piston-cylinder device. It operates on either a two-stroke or four-stroke cycle (*see* figure); however, unlike the spark-ignition gasoline engine, the diesel engine induces only air into the combustion chamber on its intake stroke. Diesel engines are typically constructed with compression ratios in the range 14:1 to 22:1. Both two-stroke and four-stroke engine designs can be found among engines with bores (cylinder diameters) less than 600 mm (24 inches). Engines with bores of greater than 600 mm are almost exclusively two-stroke cycle systems.

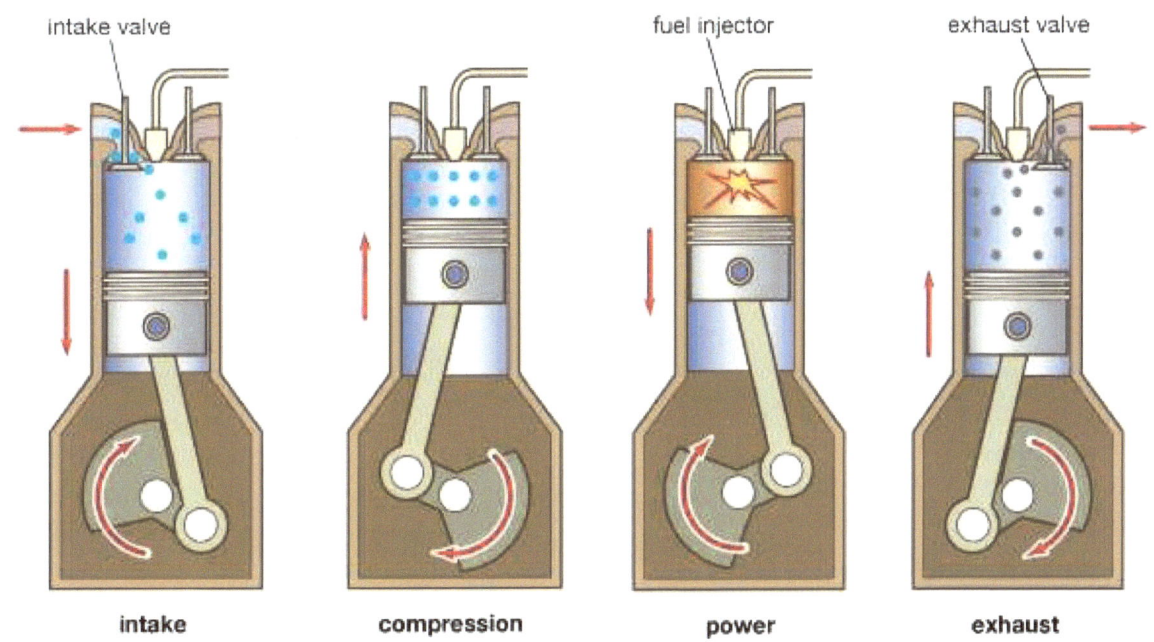

Four-stroke diesel engine

The typical sequence of cycle events involves a single intake valve, fuel-injection nozzle, and exhaust valve, as shown here. Injected fuel is ignited by its reaction to compressed hot air in the cylinder, a more efficient process than that of the spark-ignition internal-combustion engine.

The diesel engine gains its energy by burning fuel injected or sprayed into the compressed, hot air charge within the cylinder. The air must be heated to a temperature greater than the temperature at which the injected fuel can ignite. Fuel sprayed into air that has a temperature higher than the "auto-ignition" temperature of the fuel spontaneously reacts with the oxygen in the air and burns. Air temperatures are typically in excess of 526 °C (979 °F); however, at engine start-up, supplemental heating of the cylinders is sometimes employed, since the temperature of the air within the cylinders is determined by both the engine's compression ratio and its current operating temperature. Diesel engines are sometimes called compression-ignition engines because initiation of combustion relies on air heated by compression rather than on an electric spark.

In a diesel engine, fuel is introduced as the piston approaches the top dead centre of its stroke. The fuel is introduced under high pressure either into a precombustion chamber or directly into the piston-cylinder combustion chamber. With the exception of small, high-speed systems, diesel engines use direct injection.

Diesel engine fuel-injection systems are typically designed to provide injection pressures in the range of 7 to 70 megapascals (1,000 to 10,000 pounds per square inch). There are, however, a few higher-pressure systems.

Precise control of fuel injection is critical to the performance of a diesel engine. Since the entire combustion process is controlled by fuel injection, injection must begin at the correct piston position (i.e., crank angle). At first the fuel is burned in a nearly constant-volume process while the piston is near top dead centre. As the piston moves away from this position, fuel injection is continued, and the combustion process then appears as a nearly constant-pressure process.

The combustion process in a diesel engine is heterogeneous—that is, the fuel and air are not premixed prior to initiation of combustion. Consequently, rapid vaporization and mixing of fuel in air is very important to thorough burning of the injected fuel. This places much emphasis on injector nozzle design, especially in direct-injection engines.

Engine work is obtained during the power stroke. The power stroke includes both the constant-pressure process during combustion and the expansion of the hot products of combustion after fuel injection ceases.

Diesel engines are often turbocharged and after cooled. Addition of a turbocharger and after cooler can enhance the performance of a diesel engine in terms of both power and efficiency. The most outstanding feature of the diesel engine is its efficiency. By compressing air rather than using an air-fuel mixture, the diesel engine is not limited by the preignition problems that plague high-compression spark-ignition engines. Thus, higher compression ratios can be achieved with diesel engines than with the spark-ignition variety; commensurately, higher theoretical cycle efficiencies, when compared with the latter, can often be realized. It should be noted that for a given compression ratio the theoretical efficiency of the spark-ignition engine is greater than that of the compression-ignition engine; however, in practice it is possible to operate compression-ignition engines at compression ratios high enough to produce efficiencies greater than those attainable with spark-ignition systems. Furthermore, diesel engines do not rely on throttling the intake mixture to control power. As such, the idling and reduced-power efficiency of the diesel is far superior to that of the spark-ignition engine.

The principal drawback of diesel engines is their emission of air pollutants. These engines typically discharge high levels of particulate matter (soot), reactive nitrogen compounds (commonly designated NOx), and odour compared with spark-ignition engines. Consequently, in the small-engine category, consumer acceptance is low.

A diesel engine is started by driving it from some external power source until conditions have been established under which the engine can run by its own power. The simplest starting method is to admit air from a high-pressure source—about 1.7 to nearly 2.4 megapascals—to each of the cylinders in turn on their normal firing stroke. The compressed air becomes heated sufficiently to ignite the fuel. Other starting methods involve auxiliary equipment and include admitting blasts of compressed air to an air-activated motor geared to rotate a large engine's flywheel; supplying electric current to an electric starting motor, similarly geared to the engine flywheel; and applying a small gasoline engine geared to the engine flywheel. The selection of the most suitable starting method depends on the physical size of the engine to be started, the nature of the connected load, and whether or not the load can be disconnected during starting.

Types of Diesel Engine:

Three basic size groups

There are three basic size groups of diesel engines based on power—small, medium, and large. The small engines have power-output values of less than 188 kilowatts, or 252 horsepower. This is the most commonly produced diesel engine type. These engines are used in automobiles, light trucks, and some agricultural and construction applications and as small stationary electrical-power generators (such as those on pleasure craft) and as mechanical drives. They are typically direct-injection, in-line, four- or six-cylinder engines. Many are turbocharged with after coolers.

Medium engines have power capacities ranging from 188 to 750 kilowatts, or 252 to 1,006 horsepower. The majority of these engines are used in heavy-duty trucks. They are usually direct-injection, in-line, six-cylinder turbocharged and after cooled engines. Some V-8 and V-12 engines also belong to this size group.

Large diesel engines have power ratings in excess of 750 kilowatts. These unique engines are used for marine, locomotive, and mechanical drive applications and for electrical-power generation. In most cases they are direct-injection, turbocharged and after cooled systems. They may operate at as low as 500 revolutions per minute when reliability and durability are critical.

Two-stroke and four-stroke engines
As noted earlier, diesel engines are designed to operate on either the two- or four-stroke cycle. In the typical four-stroke-cycle engine, the intake and exhaust valves and the fuel-injection nozzle are located in the cylinder head (*see* figure). Often, dual valve arrangements—two intake and two exhaust valves—are employed.

Use of the two-stroke cycle can eliminate the need for one or both valves in the engine design. Scavenging and intake air is usually provided through ports in the cylinder liner. Exhaust can be either through valves located in the cylinder head or through ports in the cylinder liner. Engine construction is simplified when using a port design instead of one requiring exhaust valves.

Development of Diesel Engine:

Early work

Rudolf Diesel, a German engineer, conceived the idea for the engine that now bears his name after he had sought a device to increase the efficiency of the Otto engine (the first four-stroke-cycle engine, built by the 19th-century German engineer Nikolas Otto). Diesel realized that the electric ignition process of the gasoline engine could be eliminated if, during the compression stroke of a piston-cylinder device, compression could heat air to a temperature higher than the auto-ignition temperature of a given fuel. Diesel proposed such a cycle in his patents of 1892 and 1893.

Originally, either powdered coal or liquid petroleum was proposed as fuel. Diesel saw powdered coal, a by-product of the Saar coal mines, as a readily available fuel. Compressed air was to be used to introduce coal dust into the engine cylinder; however, controlling the rate of coal injection was difficult, and, after the experimental engine was destroyed by an explosion, Diesel turned to liquid petroleum. He continued to introduce the fuel into the engine with compressed air.

The first commercial engine built on Diesel's patents was installed in St. Louis, Mo., by Adolphus Busch, a brewer who had seen one on display at an exposition in Munich and had purchased a license from Diesel for the manufacture and sale of the engine in the United States and Canada. The engine operated successfully for years and was the forerunner of the Busch-Sulzer engine that powered many submarines of the U.S. Navy in World War I. Another diesel engine used for the same purpose was the Nelseco, built by the New London Ship and Engine Company in Groton, Conn.

The diesel engine became the primary power plant for submarines during World War I. It was not only economical in the use of fuel but also proved reliable under wartime conditions. Diesel fuel, less volatile than gasoline, was more safely stored and handled.

At the end of the war many men who had operated diesels were looking for peacetime jobs. Manufacturers began to adapt diesels for the peacetime economy. One modification was the development of the so-called semidiesel that operated on a two-stroke cycle at a lower compression pressure and made use of a hot bulb or tube to ignite the fuel charge. These changes resulted in an engine less expensive to build and maintain.

Price's engine:

In 1914 a young American engineer, William T. Price, began to experiment with an engine that would operate with a lower compression ratio than that of the diesel and at the same time would not require either hot bulbs or tubes. As soon as his experiments began to show promise, he applied for patents.

In Price's engine the selected compression pressure of nearly 1.4 megapascals (203 pounds per square inch) did not provide a high enough temperature to ignite the fuel charge when starting. Ignition was accomplished by a fine wire coil in the combustion chamber. Nichrome wire was used for this because it could easily be heated to incandescence when an electric current was passed through it. The experimental engine had a single horizontal cylinder with a bore of 43 cm (17 inches) and a stroke (maximum piston movement) of 48 cm (19 inches) and operated at 257 revolutions per minute. Because the nichrome wire required frequent replacement, the compression pressure was raised to 2.4 megapascals (348 pounds per square inch), which did provide a temperature high enough for ignition when starting. Some of the fuel charge was injected before the end of the compression stroke in an effort to increase the cycle timing and to keep the nichrome wire glowing hot.

In the meantime many engines of the two-stroke-cycle, semidiesel type were being installed. Some were used to produce electricity for small municipalities, while others were installed in water-pumping plants. Many provided power for tugs, fishing boats, trawlers, and workboats.

In the early 1920s the General Electric Company suggested to the Ingersoll-Rand Company, for whom Price was working, that they cooperate in the building of a diesel-electric locomotive. At that time many of the locomotives in service were powered by gasoline engines. A diesel-electric locomotive with Price's engine was completed in 1924 and placed in service for switching purposes in New York City. The success of this locomotive resulted in orders from railroads, factories, and open-pit mines. The engine used in most of these installations was a six-cylinder, 25-cm (10-inch) bore, 30-cm (12-inch) stroke system, rated 300 brake horsepower at 600 revolutions and weighing 6,800 kg (15,000 pounds).

End of Book

www.ingramcontent.com/pod-product-compliance
Lightning Source LLC
Chambersburg PA
CBHW051939210526
45473CB00006B/2312